Slide-out rooms mechanics and Structures

Dave Galey

Published by:
WINLOCK GALEY
26135 Murrieta Road
Sun City, CA 92585

Slide-out rooms mechanics and Structures

ISBN: 1-890461-11-3

Revised: 8-21-98

Slide-out rooms, mechanics and Structures

Slide-out rooms mechanics and Structures

Dave Galey

Contents

Warning-Disclaimer

This book is designed to provide information only on the subject matter covered. It is sold with the understanding that the publisher and author are not engaged in rendering legal, accounting, engineering, or other professional services. If legal or other expert assistance is required, the services of a competent professional should be sought.

It is not the purpose of this manual to reprint all the information otherwise available to the author and/or publisher, but to complement, amplify and supplement other texts. You are urged to read all the available material, learn as much as possible about bus conversion and to tailor the information to your individual needs.

Every effort has been made to make this manual as complete and as accurate as possible. However, there may be mistakes both typographical and in content. Therefore, this text

.

should be used only as a general guide and not as the ultimate source of slide-out room design and fabrication for bus conversions. Furthermore, this manual contains information on bus conversion only up to the printing date.

The purpose of this manual is to educate and entertain. The author and **WINLOCK GALEY** shall have neither liability nor responsibility to any person or entity with respect to any loss or damage caused, or alleged to be caused, directly or indirectly by the information contained in this book.

If you do not wish to be bound by the above, you may return this book to the publisher for a full refund.

Slide-out rooms, mechanics and Structures

Slide-out rooms mechanics and Structures

Dave Galey

Introduction

The slide out room in a recreational vehicle expands the living space immensely. A small expansion can yield the difference between cramped quarters and cozy comfort. Building on the success of *The Bus Converter's Bible* inspired me to elaborate on the Extendable Floor Space chapter with the idea of presenting the various options to create such a model. Not only do we examine the various structural problems but present several solution to them. In addition, detailed drawing are provided and optional actuations methods are given. Furthermore, consideration is given to the problems of sealing and latching.

Cutting out a large window in the side of vehicle is fraught with perils. It is the purpose of this discussion to show ways to minimize these perils and create a structurally sound unit which will take the occupants over the most interesting terrain with safety and comfort.

It is also the intension of this writer to nudge the imagination of the self conversion specialist to bring forth a coach with a slide-out room for every purpose. Consider a bus with a Living room couch slide-out, one for the computer center, a dining area expansion and a bedroom extension. In ad-

.

dition, on the top of the coach is a railed patio with a sun shade tent and the baggage compartment contains a spare bedroom, complete with it's own windows, TV and bed. To the average attendee at Indy, this rig looks like a Russian space station. When the show is over, however, everything collapses into itself and again becomes the sleek behemoth of the Interstate.

Since we have no specific motorcoach brand, or make in mind when presenting the enclosed material, the information is given in the form of concepts and ideas. Anyone applying this information to a specific slide-out room size, will have to adapt the ideas shown to their make of coach. In some respects, from necessity, the drawings are general in nature. It is hope enough detail is shown so the reader may adapt some of the concepts to their plans. The actual structure of the slide-out room is purposely left vague since the dimensions will vary according to it's use. Keep in mind,

Slide-out rooms, mechanics and Structures

a slide-out room is not much more than an oversized drawer. As far as the attachment of such things as the slide mechanism, the seals and the latches, these detail are again, purposely omitted since the actual details will vary according to the make and model of the bus.

A lot of elementarily structural analysis is presented to show the care and thought given to the problem. Even though the approach is somewhat simplistic, the author welcomes any criticism as to his approach and conclusions. In no way does the author claim the ideas shown are the penultimate. For the most part, the concepts shown are presented such that they may be done by the average skillful, self-conversion-specialist. Another way of saying that, is: *To achieve the goals of this material, you don't need a factory.*

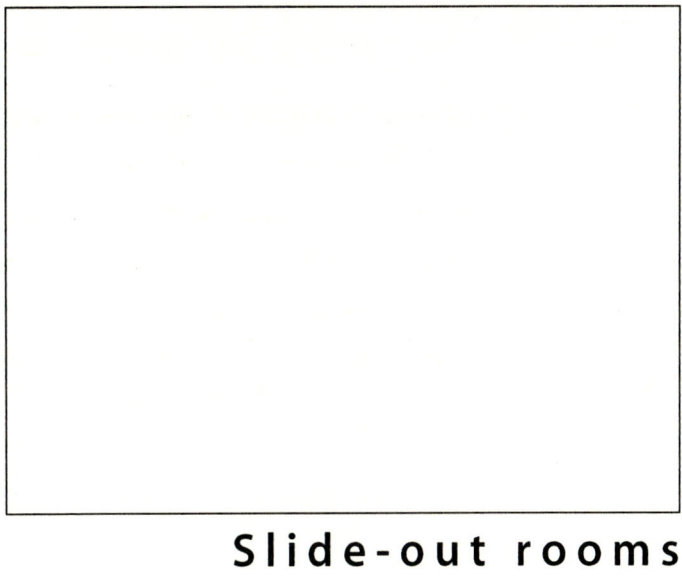

Slide-out rooms mechanics and Structures

Dave Galey

Structural Study

Lets examine the problem created by chopping a huge window out of the side of our bus. Obviously, you'll remark, "Who needs a window? I want a slide-out!" Okay! I could call the opening the slide-out will occupy a fenestration, but at any rate, it's a hole! (Editor's note: the writer is already getting tired of writing slide-out and had asked if it is okay to use the acronym SO, so OK it'll be SO {later he may want to use PU for pop-up}).

The body of a bus experiences three major forms of stress. These are bending, shear and torsion. The strain created by these three types of stress are sag, displacement and twist. Just for fun, lets decide here and now that we don't want any sag, displacement or twist. Actually a little of these strains cannot be avoided, but as long as they take place within the elastic limit of the material, there is no harm done. Too much movement (or strain) of the bus body due to these stresses can cause the bus to behave badly. What might that bad behavior be? It can cause misalignment, wearing expensive tires unnecessarily or make the bus continuously try for the other lane or even make the owner alter his equilibrium when walking from the front to the back when the bus is parked.

So we can now think of a bus body as a big box beam resting on an elastic foundation. The foundation being such things as springs, air bags, torsion bars or under inflated tires. The loads this big box beam are subject to, are dead loads (the weight of the material of which the bus is made), live loads (the weight of the cargo and/or the people being carried by the bus) and the dynamic loads, (those loads caused by driving over Arizona and West Virginia roads {my apologies to Senator Byrd})

We'll begin with the dead load of the bus, i.e., the weight of the empty shell. Under this condition, we also included the weight of the fuel and the water. This is known as the *wet* condition. In a motorhome application, wet also included the potable water but not the waste tanks; the assumption being the fresh water ultimately converts to waste water.

To simplify our problem, I choose to use a figure of 30,000 pounds for a forty foot bus and 25,000 pounds for a 35 foot bus as the dead loads respectively. This yields a uniform load of 750 pounds pet foot. A further examination of the assumptions in detail show that the average 40 footer actually weighs around 26,500 pounds empty and will hold about 1,500 pounds of liquid (fuel, water and oil), so our assumption is conservative but within a five percent error.

Most buses have a GVWR (gross vehicle weight rating) of about 1000 pounds per foot. For example, the Silver Eagle has a GVWR of 39,200 pounds, or a rating of 980 pounds per foot. This is within 98 percent of our value. Now, going back to our dead load figure of 750 lbs/ft, we see that the average live load is about 250 lbs/ft.

Finally, we use a dynamic loading factor of two and an impact factor of two giving us a total load factor of four times the combination of the dead and live loads. Such things as wind, snow and seismic loading are significant in building construction, but they are meaningless in vehicle design. The dynamic loading will be accommodated by the safety factors applied as working stresses selected. So, for the total design load, we will use 1,000 pounds per foot as shown below.

What do we do about twisting loads, and how are they applied? To simplify let us figure the bus has four corners so one-fourth of total design load may be applied at one corner. For our 40 footer, this would be 10,000 pounds. The worst condition would be three corners obeying the laws of

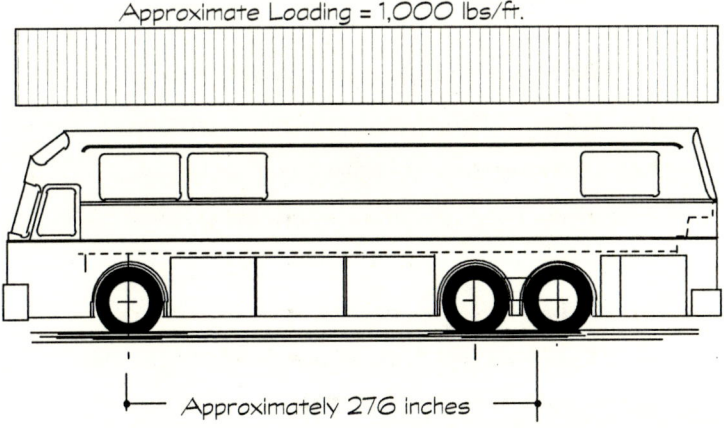

Approximate Loading = 1,000 lbs/ft.

Approximately 276 inches

gravity while one corner impacts a chuck hole, or a bump, reversing the loading. This would be our most severe torsional condition.

Now we must decide how to treat these loads. More specifically, what is the span of this box beam we call a bus? The span is what we call the wheel base. For most forty-footers the wheel base averages about 23 feet while most 35 footers use about 20 feet. (Note: for those readers accustomed to using the metric system divide by 3.1 to get meters, or .0031 to get millimeters) {Author's note: I hate the metric system; it's so damned complicated!}

At this point, I plan to introduce some terms which will be of no interest to the average reader, so if you are not interested in structural analysis, (and most of us aren't), skip to the results and conclusions.

Each shape has a series of characteristics which causes it to behave in a particular way when loaded. These

.

characteristics are known as the mechanical properties. First we have the cross sectional area. Next is the Bending Moment of Inertia, then the Section Modulus and finally the Radius of Gyration. Furthermore, these properties vary according to their axis relative to the shape. One additional property is the Torsional Moment of Inertia which we use when checking the torsional rigidity.

In simple structural problems such as floor joists and roof rafters, it is easy to pick out the size you need from tables or quickly calculate the requirements. In complex structure, we estimate the size of the parts we need and develop the mechanical properties then test to see if it works. Often this is done repeatedly until we have a satisfactory solution. This is know as iteration. Basically, it is cut and try; over-designing, then under-designing and finally zeroing in on the optimum solution. This is often, a very time consuming procedure but with today's computers, weeks have been reduced to hours.

In our quest for the answers, we have the advantage of a perfectly workable bus body carefully designed by a cadre of experts and proven over a period of years in service. It is my contention if we chop up our perfectly good bus body to insert slide-outs, we should restore it's original stiffness, hence we now have the purpose of this structural exercise.

Since our bus body is symmetrical about it's centerline, we will look at only one side to determine it's stiffness. For further simplification let us see if the bending section from the floor line to the window sill will resist the load we have selected. Since we have decided to look at only one-half of the body, we will use only one-half of the load. Let us use a span of 23 feet, or 276 inches and a section height of 30 inches. The Bending moment is:

.

$$M_{max} = wl^2 \div 8 = (500 \div 12) \times 276 \times 276 \div 8 = \textbf{396,000} \text{ inch-pounds ("#")}$$

Although the Moment of Inertia may be determined, we simplify the approach by reacting all the loading in direct stress at the extreme ends of the bending member. With a section height of 30" a direct stress would equal $396000 \div 30 = \textbf{13,200}$ pounds. This figure may be either compression or tension depending on which side of the section.

A section of steel tubing measuring 1½" x 1½" with a wall thickness of .125 " has a cross sectional area of **0.69** square inches. Dividing this area into the compression or tension load of the top and bottom members of the section give us a unit stress of:

$$f_b = 13,225 \div /0.69 = \textbf{19,166}$$
pounds per square inch (psi)

Slide-out rooms, mechanics and Structures

This number is considered the most common maximum working stress for mild steel such as cold rolled SAE 1020, or 1018. The actual yield stress for mild steel is in the 60,000-65,000 psi range. However, to yield means a permanent stretch or collapse and this is not a good thing. This little exercise, however, has demonstrated the level of safety built into the average bus. Since we applied a load factor of four and found even this has a working factor of three before catastrophic failure, it shows we can do a lot of things to our bus body without really screwing up too badly.

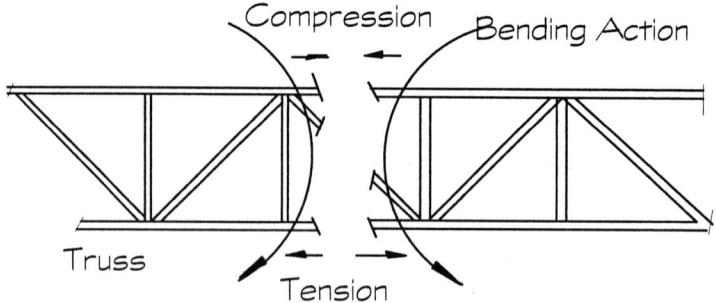

What was demonstrated was the most simplistic form of stress analysis. Actually, the loading and the stresses are much more complex, but in addition, there is much more redundancy. So, if one section fails another takes over and does it's job. Probably one of the most complex issue in a structural frame are the joints. Often in analysis we consider a joint to be pinned whereas a welded rigid joint will exhibit far more strength.

Shear loading (V) is next on the agenda. A simplified shear picture is one-half of the load per foot over the clear span. This becomes:

$$V = wl \div 2 = 500 \times 23 \div 2 = 5,750$$
pounds

To equate this to a unit shear stress, we have the area of the upper and lower chords of our bending section which are twice the cross sectional area of the square tubing noted above. The numbers are:

Slide-out rooms, mechanics and Structures

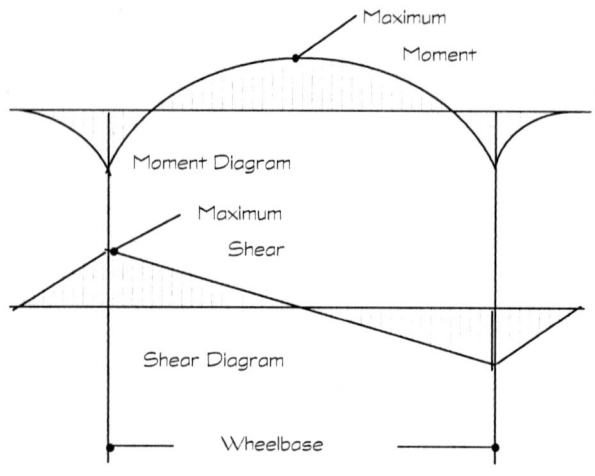

Maximum
Moment

Moment Diagram

Maximum
Shear

Shear Diagram

Wheelbase

v (unit shear stress) = 5.750 ÷ 2 x 0.69 = 4,166 psi.

This figure is far below the maximum allowable shear stress of 15,000 psi for mild steel. For all practical purposes we may dismiss shear stresses contributing to any significance.

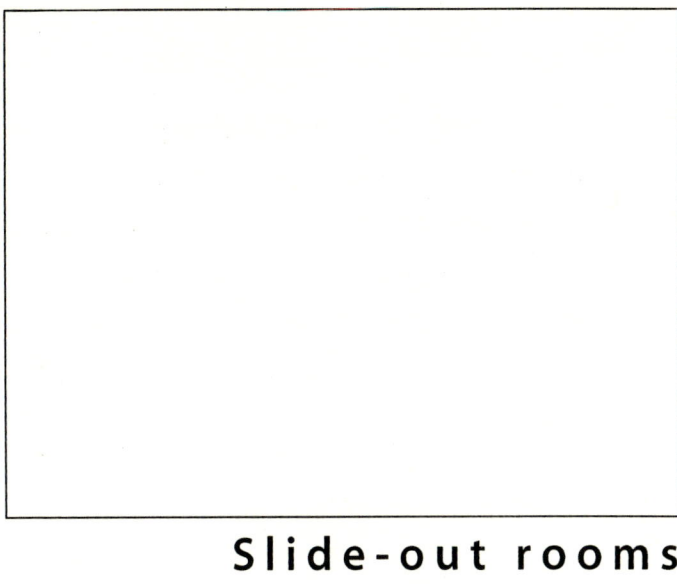

Dave Galey

Selecting Sizes

Our approach to solving the stiffness problem is to design enough structure to double around the opening so we have both the equivalent bending, shear and torsional resistance. Since we are extending the framework further apart than the old truss-work chords, bending shouldn't be a problem. Actually the bending load can be solved by treating the upper chord replacement frame as a horizontal compression member and solving it as a slender column. The torsional resistance will be a little more difficult since we have to consider the twisting of the replacement frame.

It is my intention to make some broad assumptions and solve for a replacement frame for the following dimensions and locations:

1. A living room/dining area slide-out measuring 12 feet in length and 66 inches high.

2. A couch/end table slide-out measuring 114 inches in length and 54 inches high.

3. A dining area slide-out measuring 74 inches wide and 66 inches high.

4. A computer center measuring 40 inches wide and 54 inches high.

5. A queen-size bed measuring 96 inches wide and 48 inches high.

The slide dimension for each of the above list projects will range from

24 inches to 34 inches. This is not an important consideration at this time when examining the body structure. Later, when we begin dimensioning and determining mechanisms the length of extension and retraction will become very important.

Let us begin by starting with the smallest opening in our side which would be a computer center and then working out way up to the largest opening, the combined living/dining area. The computer center need only extend 24 inches, so we will estimate the weight of the extension for mechanical support. With the weight of the user being supported by the original floor all we need concern ourselves with is the weight of the desk, the slide structure and the equipment. The slide structure should weight about 40 pounds per foot, or about 150 lbs. The desk is 60 lbs and the computer equipment is no more than 40 lbs. This gives us a total slide-out *drawer* of about 250 lbs.

Slide-out rooms, mechanics and structures

Now, lets investigate the cutout of the bus body frame necessary to accept our computer center. If we locate the center over one of the axle, we have virtually no bending loads. Still, lets us examine the worst case which would be in the middle of the wheelbase where maximum bending loads exist. One other restriction should be to limit our new structure to the thickness of the original structure. That is, we wish to keep our design with the original contour. Referring to the numbers mentioned earlier, the maximum bending moment is 396 in-kips. With an opening of 54 inches in height, we have a direct compression load of 396÷54 = 7,333 lbs.

We now have to treat this upper frame member as a horizontal column. Lets try a 1½ x 2 x 11 ga rectangular tube. The area is .782 so the compression stress is 7,333 ÷ .782 = 9,377 psi. This is no problem unless we have column instability, or buckling. The

.

formula for elastic instability for this type of section is:

$$P \div A = s = 25{,}000 - .425(L \div r)^2$$

For our application , s = 25,000 - 3610 = 21390. This mean we are well within to limits of column stability by a factor of more than two.

Our next approach is to investigate the load transfer from the penetrated bending section to the vertical members of our frame. In this case we have a 13,225 lb compression load intersecting our vertical member at 30 inches above the floor line. This gives us a bending moment of 7,333 x 24 = 175,990 in-lbs. Again we will assume a 1½ x 2 in rectangular tube and check for bending stresses. The Section Modulus of this member is .428. Therefore 175,990 ÷ .428 = 411,190 psi. this is obviously inadequate for this application. To keep our bending stresses down to or below 20,000 psi, we need a Section modulus of 175,990 ÷ 20,000

= 8.8, or better. A Section Modulus of this value requires at least an eight inch channel with 2½ inch flanges. A simpler approach would be to add a diagonal brace to this vertical member, thus creating a small truss section which would be in direct stress.

The logical approach would be to carry the diagonal member over to the closest point on the existing truss as shown in the drawing below. Assuming diagonal with a 30 degree included angle, would then have a direct compression load of 7,333 x sin 30 = 14,666 lbs, or a stress of 14,666 ÷ .782 = 18,754 psi. this is well within acceptable limits . If the coach is designed as a monocoque structure, it would be

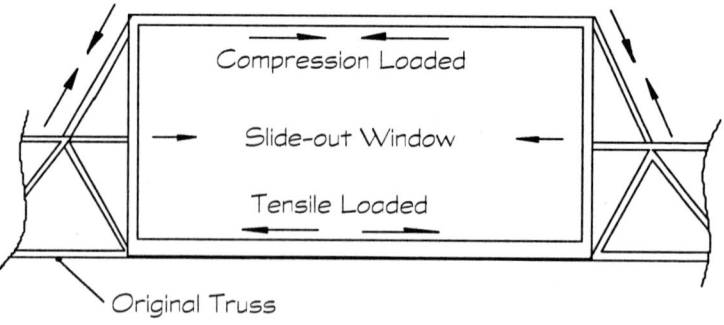

Compression Loaded

Slide-out Window

Tensile Loaded

Original Truss

appropriate to add a set of diagonals. that is to duplicate the shape shown on page 17, making sure to extend a diagonal from the header to the sill and then from the sill to the floor on each side of the opening.

Next, we must investigate the bending capability of the lower frame member. We may use the bending moment formula for a fixed end beam: $M = wl^2 \div 12$. In this case W = 500 lbs/ft., or 41.67 lbs/inch. Lets use the maximum opening length for the section we examined above. Therefore, the bending moment would be, M = 41.67 x 85 x 85 ÷ 12 = 25,090 in=lbs. The bending stress would equal, $f_b = M \times c \div I = 25{,}090 \times .94 \div 0.505 = 46{,}702$ psi. This value is too high for the working stress of mild steel, so we will try another section. I value for a 1.5 x 3 x .125 rectangular section is 1.34. The Section modulus is 1.34 ÷ 1.50 = .8933 in-sq.. So. The extreme bending stress for this section over an 85 inch span are 25,090 ÷ .8933 = 28,090 psi. Although

slide-out rooms, mechanics and Structures

this is marginal, we have enough fear factors (safety margins) built in so this section would be acceptable. We now have the stiffening frame for the smaller of the windows with 1½ x 2 inch section up the sides and across the top, with a 1½ x 3 inch section for the lower member. As an added stiffness at the joint we will weld in a gusset which will also serve as an indexing position for the retracted slide-out. This means we will have indexing holes drilled into these gussets which will accept alignment pins location in the movable section.

Our next test will be to size the lower member for the larger of the two slide-out windows. Again the formula for the bending moment is $M = wl^2 \div 12..$ And again the loading will be 41.67 pounds per inch. With a maximum opening size of 13'-5" the bending moment is 90,010 in-lbs. This time we will solve for the required section modulus and select the section from tables. Allowing for a maximum bending stress for mild steel of 25,000 psi, the section modulus will be 90,010 ÷ 25,000 = 3.6 in-sq. A 2" x 4" rectangular tube with a wall thickness of .125 is marginal, however this section with a 3/16 wall thickness is more than adequate. If the opening width is limited to 12 feet, then the one-eighth inch thick wall would serve quite satisfactory. So, now we have sized our large frame for our slide-out window at 2" x 3" x 0.10" rectangular section for the sides and upper members with a 2" x 4" x 0.125" rectangular section for the lower frame.

This analysis more or less defines the lower practical limit of a slide-out opening. Next we will examine the upper practical limit. Although some openings have been as large as 15 feet, my personal feeling toward the practical upper limit is 12 feet. The guiding factor of the 12 foot limit is again elastic instability of the compression member replacing the upper chord in the truss section. Again we need to react a bending moment of 396,000 in-lbs. Only, this time we are using an opening height of 66 inches. 396.000 ÷ 66 = 6,000 lbs of direct compression. The column formula to avoid buckling is $P \div A = 25{,}000 - .425\{L \div r\}^2$ The expression, $l \div r$, is known as the slenderness ratio. Obviously, the more slender a column is, the less resistance it has to buckling. So, lets first solve to the slenderness ratio of a 12 foot opening with the same rectangular tube we selected previously: $L = 144"$ $r = 0.586"$ (note: the technical name for "r" is radius of gyration). The radius of gyration is defined as the distance from the centroid of the section where all the mass may be concentrated relative to the axis in question. For column design we always select the least radius of gyration. Therefore: $144 \div .586 = 244$ in^2. $244 \times 244 \times .425 > 25{,}000$. Consequently $P \div A = 25{,}000 - 25{,}000+ = <=$ zero. This section will not be stable for our loading, so we must select another

.

member with more stiffness to span this opening.

Tentatively we select a rectangular section 2 x 3 x 14 ga. The member has a radius of gyration value of 0.823. Solving the above equation gives us a stress capability of 11,900 psi. with a cross sectional area of 0.802 inches and a compression load of 6,000 pounds, we have a stress level of 7.480 psi. this gives is a safety factor of 1.5.

The cross over point of the 1½ x 2 section may now be solved using the column formula and working backwards:

$$6000 \div .782 = 7672 \text{ psi. } 7672 =$$
$$25,000 - .425\{L \div r\}^2 \quad L = .425 \times 40770$$
$$= .425 \times 201$$

L = 85.4 inches. This means we can use the smaller section up to a maximum opening of about seven feet. In excess of that length we must use a stiffer section or we will experience a buckling failure. Actually, considering the location, the smaller section is acceptable for the 90 inch bed opening since we are in an area of minimum bending. Now that we have established the maximum width of opening for the smaller section, lets now establish the maximum length for the larger section. The stress will be the same. The only variable is the r value. $L = .802 \times 201 = 161$ inches. This is 13 feet 5 inches, which I feel is the maximum practical limit.

Again, we have the side member bending which must be reacted. In this case, our bending moment is $6{,}000 \times 36 = 216{,}000$ in-lbs. As noted above, to react this degree of bending with a structural member would be essessively heavy and difficult to achieve, so again, we look to the diagonal truss member for our solution. To convert this value to a 60 degree diagonal load, we again have $6{,}000 \times \sin 30 = 12{,}000$ pounds load in direct compression. The length of the diago-

·

nal would be 36 ÷ cos 30 = 41.5 inches. P/A = 25,000-.425(L/r)^2. = 22,864. Actual compression stress is 6,000÷0.782 = 7,680 psi (this is again well within the acceptable compression stress limits.)

Shear is the next form of stress to investigate. The maximum value of shearing load would be one-half of one half of the total weight of the bus. Assuming a total weight of 40,000 pounds, we use 10,000 for the maximum shear load. The position of this maximum load we assume to be at either end of the opening. In addition, the material available to react this loading is the top and bottom members of the opening frame. In the case of the smaller section, we have the formula:

V = 10,000   v = 10,000 ÷ 2 x 0.782 = 6,400 psi (this is well within the acceptable limits for shear stress).

We now need to examine one more problem concerning the stiffness frame. If you will look at the drawing on page 42, it may be seen that the lower structural member is inset from the members next to the outside contour. The reason for this inset is to permit the side on the slide-out room to extend below the floor so it may be supported by the telescoping tubes and still retract in a manner flush with the outside contour. Where this member attaches to the existing framework is a joint subject to both tension and shear. The tension load is determined to be 6,000 pound at each end. Using a weld shear value of 15,000 psi and a 3/16 inch fillet on each side of the member, we have: 6,000/(.188 x 15,000 x 2) = 1.07 inches of weld. Under these conditions, I would apply 4 inches of weld each side to account for porosity, cold shuts and other fear factors. The shear loading at this joint would be 13.8 x 500/2 = 3450 pounds, which is less than the tensile loading in shear, so the 4 inch overlap at each end of the opening is adequate.

Finally, torsional stresses must be investigated but more importantly,

.

the degree of torsional strain, or twist. Obviously, the greater the length of the opening, the more twist, or displacement we can expect, so only the largest opening will be analyzed. In this case we must look at the body in cross section as though it were an open side channel. If the body were in a moving condition with the slide-out room extended, it is doubtful we could effect a satisfactory solution be-

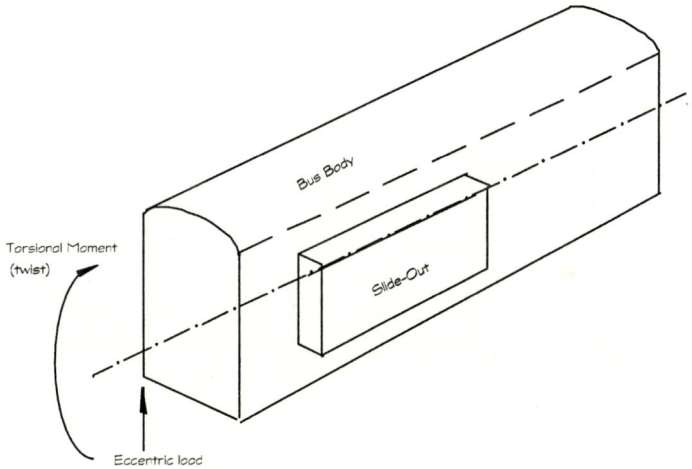

cause it would be subject to impact loading and probably twist out of shape. It is true, we could add enough stiffening to accommodate moving loads, but the excessive weight and complexity would be prohibitive.

However, for our design loads, lets us assume the bus is on an uneven surface and the leveling jacks are extended causing an eccentric torsional load condition. In a similar manner, we select the eccentric loading to be one-fourth of the total weight of the bus, or 10,000 pounds. It can be seen that excessive torsional deflection in a static condition could cause a binding, or a lockup of the extension/retraction mechanism. This would be really embarrassing if you had to drive to a service center with part of a room sticking out, not to mention the highway patrolman's ticket.

The three principle formulas we use to investigate torsion are:

(1) $s = Tr \div J$

(2) $I_p = J = A_x d^2 \times 2 + A_y d^2 \times 2$,

(3) $a = Tl \div KG$.

T = the twisting load

s = torsional shear stress

.

r = distance from the centroid to the outside fiber

a = the angle of deflection, or twist

$A_x d^2$ = a close estimate of the moment of inertia about an x axis

l = length in inches

G = the modulus of rigidity for the material (for steel, G = 11,500,000 psi)

K = shape factor (this can often times be a WAG*, *Wild Ass Guess)

Actually K is a complex calculation based on empirical data so we tend to be a little conservative.

The torsional analysis will compare a completely closed section with dimension of 66 inches high and 96 inches wide with a section with one 66 inch side removed With the enclosed section we can approximate the polar moment of inertia to determine the torsional shear stresses. However, with the open sided section we must use the least moment of inertia to determine the shear stresses.

The other assumption we use is the moment of inertia is calculated only with the skin thickness being effective. In other words, no frames or longerons will resist twisting.

The polar moment of inertia of the closed section is 2 x 96t x 33 x 33 + 2 x 66t x 48 x 48 = 30,800 inches - fourth. We assume the twisting moment to be 480,000 in=lbs, so the shear stresses are 480,000 x 48 ÷ 30,800 = 750 psi. Our soft 6061 type aluminum is quite adequate to handle such a low shear stress.

Now lets open one side of the section and check the least moment of inertia for the open section torsional shear stresses.. The least moment is 7300, so 480,000 x 33 ÷ 7300 = 2,170 psi. This is still within the limits of the material, despite the fact the stresses have gone up by a factor of nearly three.

Now lets look at the angular displacement, a little tougher analysis, After some time I arrived at a K factor for our closed section of 490,000. Our twist angle over the 161 inch distance is:

480,000 x 161 ÷ 490,000 x 11,500,000 = 0.0000137 radians, or 0.0008 degrees

OK, the K factor for out open section is: 41,140. The angular twist for our open section in 161 inches is now 480,000 x 161 ÷ 41,140 x 11,500,000 =

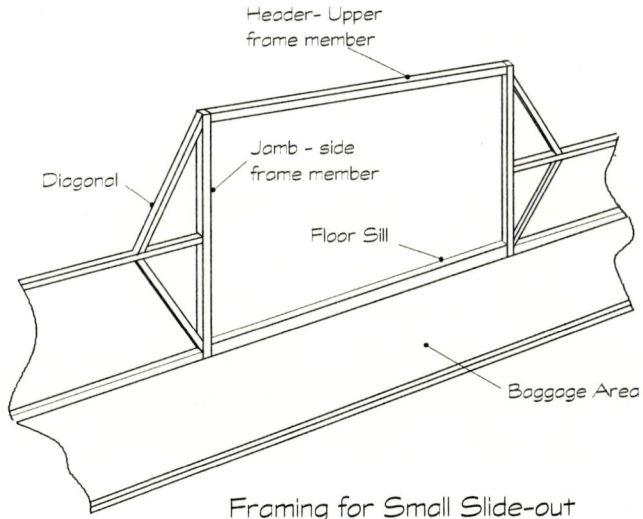

Framing for Small Slide-out

0.000164 radians = 0.01 degrees. This looks OK to me since it implies that even if we have an eccentric load con-

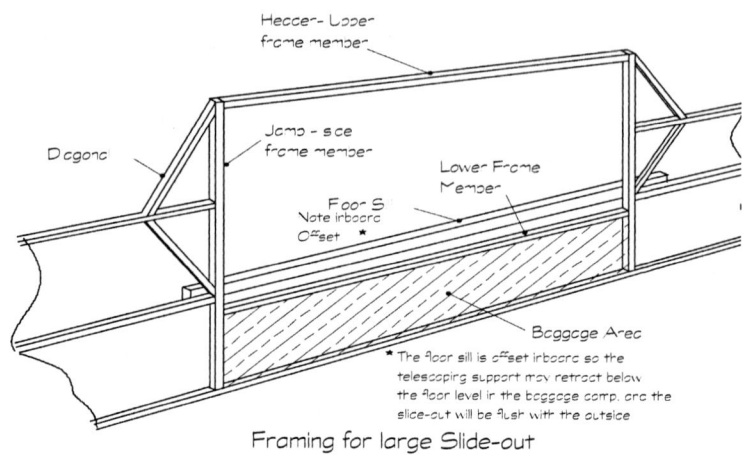

Framing for large Slide-out

dition generated by our levelers, we will only see a total twist of less than one-quarter inch (¼"). In addition, we picked the worst condition possible so, coupled with the frame design we selected earlier, we will have an adequate structure.

Our next chapter will examine the actual slide-out methods.

Summary of Frame Elements

Member	Large Frame	Small Frame
Floor Sill	2" x 4" x .188	1.5" x 3" x .125
Header	2" x 3" x .100	1.5" x 2" x .125
Jamb	2" x 3" x .100	1.5" x 2" x .125
Diagonals	2" x 2" x.100	1.5" x 1.5" x .125

slide-out rooms, mechanics and structures

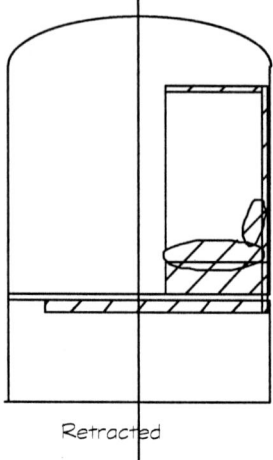

Retracted

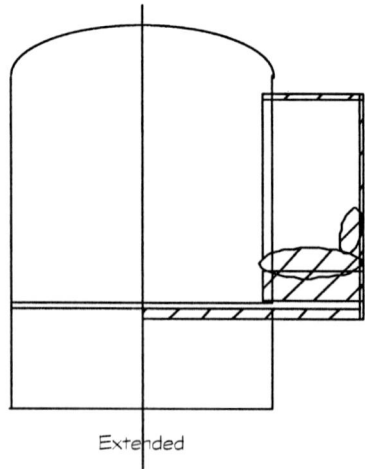

Extended

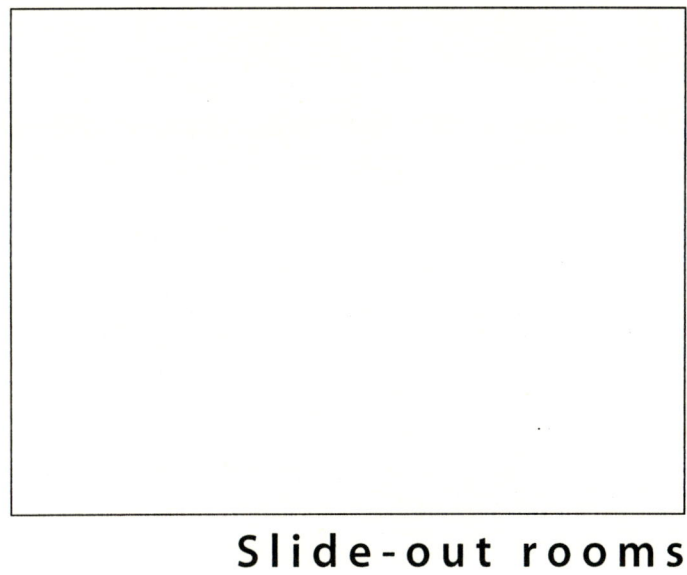

Dave Galey

The Slideout Section

Before we can decide how we are going to support the extended section, we must determine it's weight. If it is a couch section, the dead weight will be approximately 50 pounds per foot, plus an impact load of a 250 pound man, (or woman) flopping down on the couch. This impact load can achieve a momentary load of 500 pounds, so lets assume a design load of 100 pounds per foot of slide-out and we'll be conservative.

The next thing to consider is how to support this great sliding device. If it were supported entirely from within the opening, such as a drawer in a cabinet, we have one problem. On the other hand, the unit may be supported from below the floor line and this yields a different set of problems.

Choosing the smallest application, the pop-out computer center, gives us a total design load of 333 pounds. This can also be divided into to 167 pounds per side.. In such a minimum design condition, it would be possible to solve the support problem with two pair of 100 lb. Full extension 24 inch drawer slides. In other words, what we are doing is to manifold two set of drawer slides on each side of our drawer-like slide-out. In this application our design loads are quite conservative in that our desk shouldn't ex-

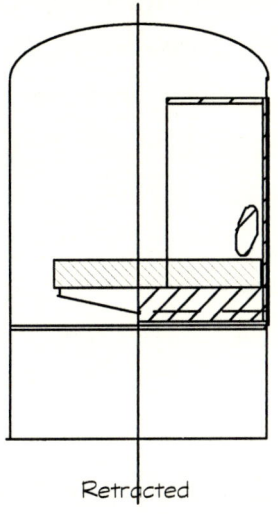

Retracted

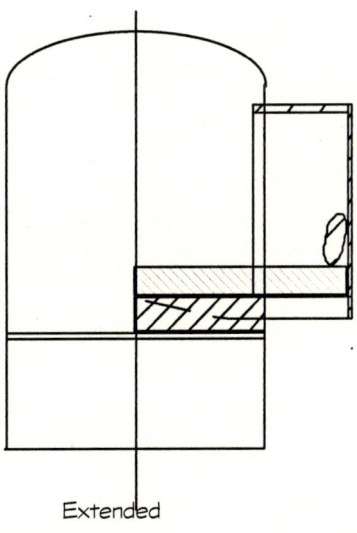

Extended

Slide-out rooms, mechanics and Structures

ceed 60 lbs, our computer, monitor and printer no more than 40 pounds and the structure is probably less than 100 pounds. And, we have the added safety that no one will flop down on the desk and the operator will be supported by the original floor structure. The hardware for this design is readily available. We simply order two sets of 100 pound rated full extension 24 inch drawer slides. These can be Accuride or Grant. The reason I select these two brands is they are of the highest quality and are ball bearing supported.

One constraint for this form of design is the necessity of maintaining the slide support structure inside of the coach. This means some form of cabinet work on either side of the moving section. This design is also applicable to the sliding bedroom extension, but has the added advantage of being able to conceal the sliding mechanism beneath the bed itself.

Although we could continue with this design philosophy for our larger slide-outs, it would become necessary to stack multiple levels of drawer slides; up to as many as ten on each side. One further limitation is most production drawer slides are available only up to 22 -24 inch extension. The design criteria for the larger slide-outs are at least the width of the normal couch. To go one step further, one might consider the width of a dinette to be the design criteria. This could require a slide extension of up to 42 inches.

For the larger units it seems necessary to support them from below the floor line. Several very effective slide support designs are available. In one application, the owner did not want to have the telescoping slides extend past the center of the coach in the baggage area so we constructed a set three piece telescoping slides extensions. Because of the tolerance buildup, this design is to be avoided, but after four years in service (at this writing), the unit has been functioning very satisfactorily. Several slide mechanism design concepts will be shown and the reader may select the one with which they feel most comfortable. It should be emphasized, no one design concept is perfect in all cases and, no doubt, better designs may be developed; perhaps by a reader.

Several points should now be brought up. First, it is desirable to have the slide-out section retract in a flush manner, much in the way a car door closes to fair with the rest of the body. In order to achieve this end, we have no problem if we are keeping the entire slide-out structure above the floor line like a cabinet drawer. On the other hand, if we extend the support structure below the floor line, we must maintain continuity from the lower section to the upper section which means we must create a structural integrity inboard of the sliding section so it may retract flush with the out side skin. The drawing on page 89 will illustrate this idea. This shows the sliding structure's outside thickness when retracted is flush with the outside contour. When the slider is extended, the body structure displays a continuity from the upper section to the lower section approximately an inch and a half inside of outside contour. You follow that?

If you are satisfied with your slider, in the retracted position, remaining outside of contour, as in many fifth-wheels and motorhomes, then it is not necessary to install this substructure. In some respects, this is a much simpler approach and a little easier to seal, since a great overlapping flange may be used with a generous gasket. Remember, the object of this treatise is to provide options, not pass judgement

The slide mechanism may be fabricated with square tubing and ball bearings, or simply a lubricated slide of metal against metal. No high speed actions is going to take place so the support slides may be very simple. One proven technique is to 2½ x 2½ x ¼ wall thickness square tubing with a 2 x 2 x ¼ inch wall thickness inner tube. The outer tube has flanges welded to one side so it may be bolted to the under side of the floor structure. The inner tube is free to telescope in and out of the outer tube. This combination will support nearly 1,000 pounds load when the inner tube is extended 30 inches.. This means if we have a 12 foot slide out with a design load of 100 lbs per foot, then two of the telescoping slides just described will be adequate to serve our needs.

As far as the sliding section is concerned, it is suggested you simply duplicate the outside wall structure and create side walls, a floor and a roof. Install any windows you wish and brace the side walls and the roof with diagonal bracing, depending on the size of the slide-out. On the slide-out floor section, plan to use wide, small diameter rollers. If a wood, or tile floor

Outside Tube: 2-1/2 x 2-1/2 x 3/16 M.T.
Inside Tube: 2 x 2 x 3/16 M.T.
Max Load @ 36" Extension: 625 LBS PER SLIDE
Max Deflection @ 36": 0.36/inch

ALIGNMENT ROLLERS

REACTION ROLLER C

POPOUT ROOM
LOADING

ALIGNMENT ROLLERS

BEARING ROLLER

is used in the bus, these rollers will work well. If you have carpet over the fixed floor, the support of the sliding section must be stiff enough to ride above, but not on, the carpet. To preserve the carpet, it may be necessary to provide rails along the side of the unit. This, however, would imply cabinet work along each side. Still, this is entirely feasible if one end of the slide-out unit was along a bulkhead and the other end by a cabinet like an end table.

With the exception of a dining area, the slide-out unit will have no exposed floor. For example, the sofa section will be occupied by the sofa. The bedroom slide-out will contain the bed and the computer center, or office, as you may prefer, will contain the desk module. In the case of the dining area, it is suggest the floor be something hard like tile or hardwood.

One other potentially unique idea for a slide-out section might be a unit containing a fireplace or a television, or both. Perhaps those hearty souls in Yukon or Nome might appreciate such an application.

As a matter of convention throughout this dissertation, 1½ x 1½

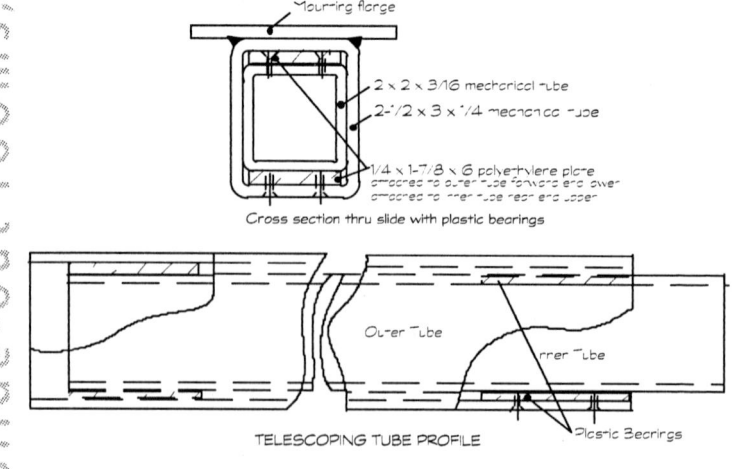

Mounting flange

2 x 2 x 3/16 mechanical tube

2-1/2 x 3 x 1/4 mechanical tube

1/4 x 1-7/8 x 6 polyethylene plate
attached to outer tube forward end lower
attached to inner tube rear end upper

Cross section thru slide with plastic bearings

Outer Tube

Inner Tube

Plastic Bearings

TELESCOPING TUBE PROFILE

inch square tubing will be used for the structure of the extendable sections. This will be true whether it is the smallest unit or the largest. This way we can develop some standards plus the fact this section can be strong enough yet small enough and weight efficient, but of a workable size to allow for insulation and some buried mechanism. The square tubing chosen for our movable sections has weight of 1.3 pounds per linear foot. So, lets determine how many feet are needed for the structure of a 3' x 6' x 12' unit. I estimate a total length of tubing to be 120 feet. This gives us a weight of structure of 156 pounds, the weight of siding is 180 square feet times 2 lbs per foot equal 360 pounds. Lets insulate and siding the interior with plywood and hardwood to give us another 250 pounds. Totaling all this weight we have almost 800 pounds. To simplify matters, lets use a weight of 66 pounds per foot. Comparing this to our origi-

slide-out rooms, mechanics and structures

nal estimate of 50 lbs/ft. dead load, we are pretty close.

If we do not use rollers to support the lower section of the slide-out and plan to have it ride above a carpeted area, then we must create enough stiffness at the support joints so the unit is cantilevered above the flooring. Using our original weight fact of 100 pounds per foot and figuring a 12 foot section extending 32 inches, our load factor is 600 pounds at 16 inches giving us a bending moment of 9,600 inch-pounds at each telescoping support member. Analyzing this for strength, we have a 4,800 pound load at each weld of the 2 inch section. Again dividing this value by the width of the support section, the unit stress is 2,400 pounds per inch divided by ¼ inch wall thickness, we have a shear stress of 9,600 psi. This is acceptable but we must be able to provide the same stiffness on the other side of the joint.

.

The other side of the joint is our sliding frame structure and we have limited ourselves to 1½ in section thickness. Looking at a 1½ inch section thickness and testing it with a 9,600 in-lbs moment, we have a bending stress of 9,600 ÷ .331 = 29,000 psi. Although this is marginal, it would work. However, the weld shear stress calculated above is no longer valid since we do not have an equivalent weld on the other side of the joint. The answer to this problem is to use 1½ x 2 x ¼ inch wall thickness members for our uprights supporting the sliding section. As additional insurance, it would be a good idea to add gusseting to these joints

We now have enough information to begin our design. Recalling our maximum size window for the sliding section, it appears to be necessary to use all those inches to achieve a practical space use. First a sofa must be at least 78 inches wide. Then we add a 14 inch end table on each side and finally a 60 inch dinette section giving us total of 166 inches, or 13'-10" wide. One editorial comment I would like to make about this type of design: the units I have inspected create a step at the dinette section and, being a clumsy oaf, I tend to stumble over this small step, not just once, but almost every time I try to sit at the table. It must be a mental defect on my part. With a great deal of complexity, we may solve this problem by allowing the sliding floor to be flush with the original floor, then having a filler floor pop-up to occupy the void after the unit is extended. Barring this level of complexity, plan to stumble over the 1½ to 2 inch step. Remember to make the opening size at least 2½ inches larger than the inside dimensions of the sliding unit. The reason for this is to provide for at least 1½ inches of structure for the sliding unit and a space for weather seals. We tentatively assign approximately 1 inch to the sealing space. The type of seal I propose is a bulb seal about three-quarters of an inch in diameter which snaps over a flange. This is the most common automotive seal available.

One more subject must be discussed before we begin our design on paper: this is actuation methods and is discussed in the next chapter.

Slide-out rooms, mechanics and Structures

Slide-out rooms
mechanics and
Structures

Dave Galey

Actuation Methods

How many methods of extend-
ing the slide-out unit can you think of?
Well, lets list a few:
1. Screw jack
2. Bell crank
3. Chain
4. Cylinder (air or hydraulic)
5. Gear drive
6. Cog belt drive

7. A good hard shove

8. Tilt the rig and let'er slide.

The screw jack is a direct action and so is the chain and the cylinder. The bell crank must be actuated with something like a cylinder or a jack since it is a push-pull process. I am personally adverse to using an air cylinder since the compressive nature of air tends to cause it to be explosive. Hydraulics is much more manageable and may be carefully controlled. Electrical systems, in general tend to be either on, or off, so it is full bore or dead stop. Electrical, however, may be slowed down to the point it is quite manageable plus you have the added advantage of being able to use limit switches.

In fact, one of the simplest form of actuation for this type of motion is a garage door opener. The unit comes with the motor, chain (or screw jack) and the limit switches. Plus, the motor is geared down slow enough to avoid impact starts and stops. Furthermore,

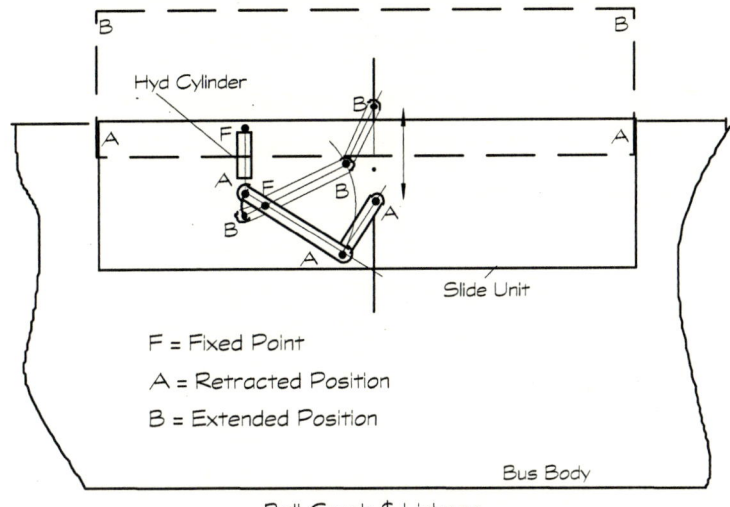

Hyd Cylinder

F = Fixed Point
A = Retracted Position
B = Extended Position

Slide Unit

Bus Body

Bell Crank & Linkage

they come with two remote control transmitters. It would be kind of weird though, if a spurious signal caused the system to actuate. This problem can be solved by using an electrical cutout when it is in the desired position. The only drawback to the use of a garage door opener is it's need for 120 volts AC. In this modern day of solid state invertors, the problem of 120 volts AC is easily solved. In addition, often you will be in an RV park or at a friends

slide-out rooms, mechanics and structures

place with house current handy. Again, this sort of decision must be based on your life style.

For those of you who must get away from it all, park by a quiet stream, 150 miles from civilization, a 12 volt DC screw jack is commonly available as a camper jack. The slide-out unit fabricated in my back yard in 1995 uses a camper screw jack. It is very slow, in that it takes 40 seconds to extend or retract. But, this allows for a low horsepower motor and it has proven to be very reliable.

Another form of actuator might be a common 12 volt DC cable winch which is easily obtainable and reasonably priced. Since a winch is a single directional device, an optional is to have one for extending and one to retracting, or to manually reposition the cable according to the direction of action desired.

Bell cranks. What are they? A bell crank is nothing more than a lever pivoting around a point being pushed and in turn pushing something else. For example a pair of bell cranks coupled and actuated simultaneously could shove a slide unit out and then retract it. Using bell cranks a movement may be amplified or attenuated. An extension of the bell cranks principle gives us the four-bar linkage which permit some really strange movements to occur such as opening a door maintaining parallelism through out the cycle.

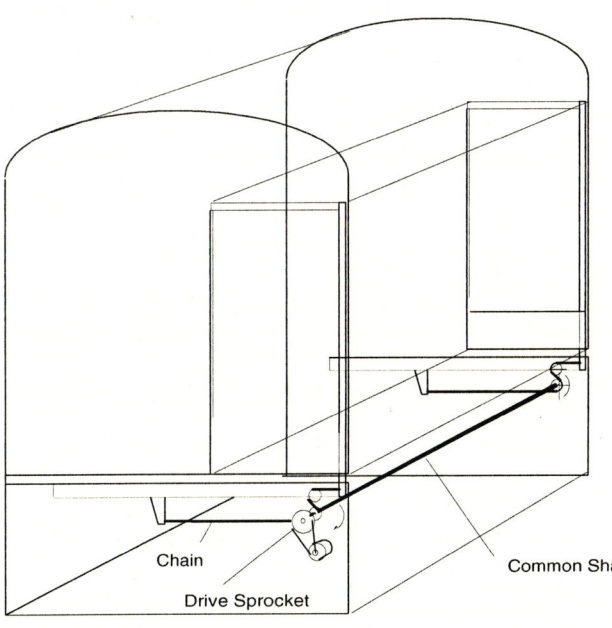

Chain

Common Shaft

Drive Sprocket

Retracted

slide-out rooms, mechanics and structures

My personal favorite would be the chain drive. With this mechanism several problems may be solve simultaneously. For example, it is important that when the sliding unit extends or retract that it does not warp, cock or bind. Several tricks can solve this problem. One way we did several years ago was to install a pair of rack and pinion gears with a common shaft. We used pinion gears about 1½ inch in diameter. These were located within a few inches of each end of the slider. On the sliding unit itself, we install a corresponding rack gear which mated with the pinions. The pinion gears were coupled with a common shaft, so that if one pinion gear rotated in it's rack, the one at the other end rotated. This would keep the sliding unit in registration and eliminated any warping, or binding due to uneven extension or retraction. Returning to the idea of the chain drive actuator, if we used sprock-

.

ets at either end of the sliding unit coupled with a common shaft, both the retraction and extension could be accomplished by the chain along with uniform movement without any binding.

The configuration of the chain system would be a pair of small sprockets designed for number 35 bicycle chain set in an "ESS" design. One of the sprocket would be a driven sprocket while the other would be an idler. With this shape, and a common shaft to another set at the other end of the unit, the chain would have the capability of both pushing and pulling. This concept is quite common with radio controlled gate openers. So this design solved both the actuation and the alignment problem. The belt drive could still be the motivating device. In fact the driving shaft could be turned by belts, chain or gears.

The idea is shown on Page 65.

Slide-out rooms, mechanics and Structures

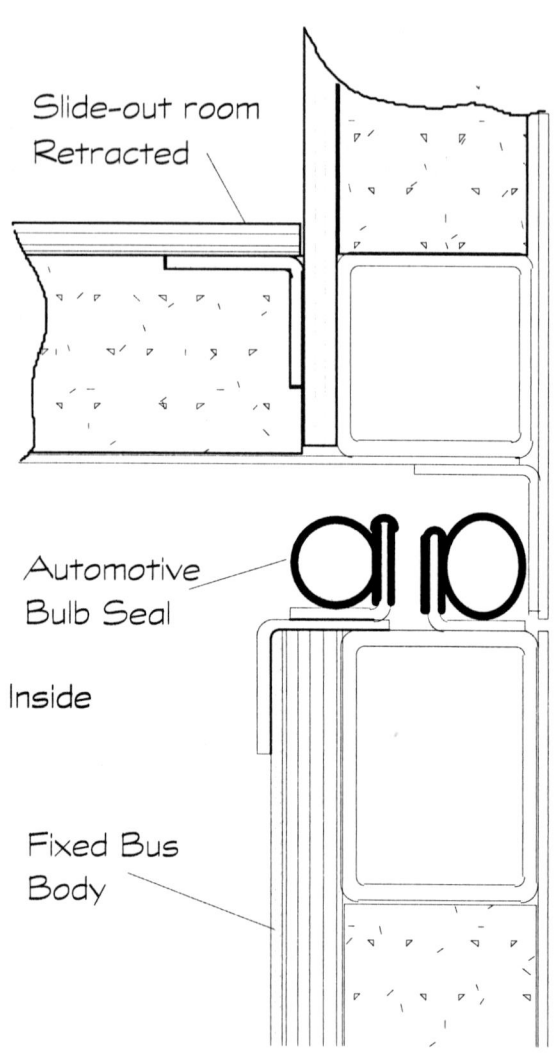

Slide-out room
Retracted

Automotive
Bulb Seal

Inside

Fixed Bus
Body

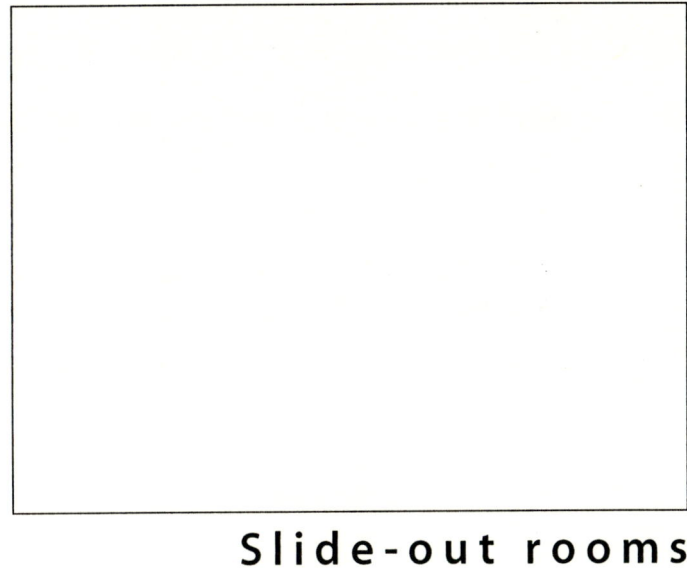

Slide-out rooms mechanics and Structures

Dave Galey

Seals

The next problem to consider is sealing. Obviously we don't want rain, wind or dust to come into our rigs. Sealing can be one of the trickiest part of the slide-out solution.

The most common automotive type seal is the bulb seal. This is a soft rubber seal about ¾ inch in diameter with about a sixteenth- inch wall thickness. It also has a metal and plastic channel attached so it will slip over a flange for retention. If you are curious to see what it looks like, open the door on a Chevy truck, and WaLaa! This type of seal is such that very little impingement is required to do it's job. With as little as an eighth inch impingement, or depression, it is effective. It is readily available at automotive upholstery supply stores.

So now we have the problem of sealing both in the retracted position and in the extended position. The simplest way to do this job is to have a pair of seals, back to back with faying surfaces on the inside extended section and on the inside retracted section. For those unfamiliar with the term "faying surface", this is the surface which touches the seal all around. In a glue joint, the faying surfaces are the overlapping surfaces. This type sealing is important for the tops and the sides. Often, it is impractical to seal the bottom in this manner. A good quality weather-stripping seal is generally satisfactory for the bottom.

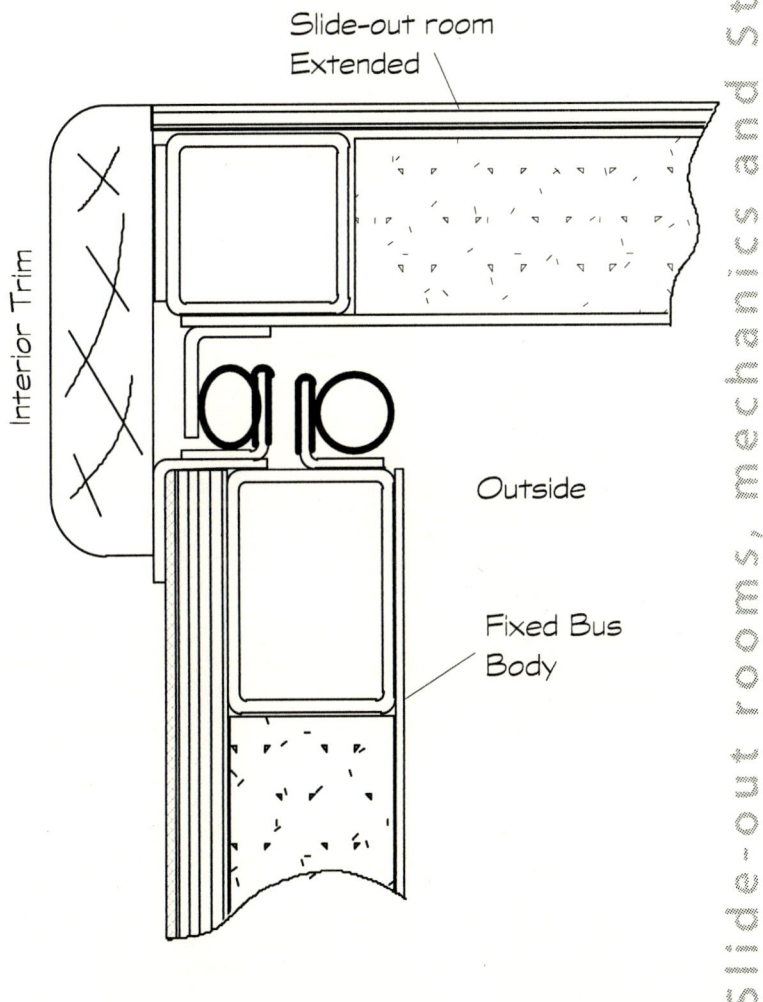

Slide-out room Extended

Interior Trim

Outside

Fixed Bus Body

As added insurance, a window awning may be attached to the slideout leaving the other side attached to the bus. then when the slide out is extended, a shading awning is extended creating a rain ans sun shade.

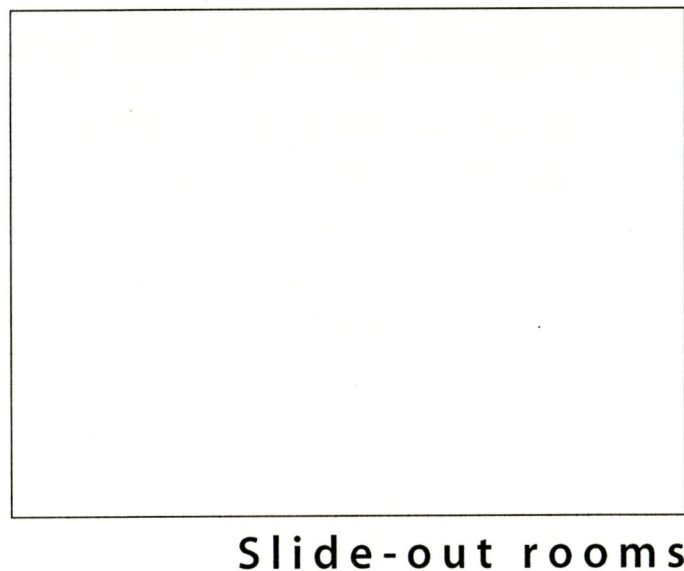

Slide-out rooms mechanics and Structures

Dave Galey

Latches

Latching the slide-out room in place when retracted is essential. It is also a good idea when the room is extended. The actuation system should not be made to maintain the room position whether extended or retracted.

Latching may be achieved at the extenders, that is the slide mechanism, or at the corners of the room. Although it is easier and simpler to latch the slide extenders, a better approach would be to lock down the outside corners of the room to the original bus body. This way, both better structural integrity and sealing is maintained.

Probably the simplest and most crude form of corner latches would be the use of "C" clamps. Another form of manual latching would be over-center De-Sta-Co tooling clamps. The over-center tooling clamp would be quite effective but might look kind of funny.

Another technique would be a simple slide bolt. The operator would slide back the bolts, extend his room then re-slide the bolts at all four corners. Although this would be a manual operation, some very attractive solid brass surface mounted slide bolts are available. The essential thing is one must not forget they are in place when attempting to extend or retract his room. This is like any check list item.

But, if one were to forget (and I am sure Murphy would guarantee one would), serious damage could occur to your actuation system.

Although failure of electrical control systems are rare,(they will fail someday), a sequencing of latching can be more reliable than the frail human brain. An example of this is the common garage door opener. This inexpensive mechanism has several failure devices built into it. It uses limit switches to disconnect, or turn off, the motor when the door is either retracted or closed. It has a clutch device to stop the motor if an obstruction is encountered. And, all new door openers now come with a photocell sender and receiver to stop the motor in the event something, such as a child or dog, breaks the beam.

It is my proposal to use solenoids to actuate slide pins for latching purposes. A simple system may be designed so that an extension mechanism switch will energize these sole-

slide-out rooms, mechanics and structures

noid latches, opening them during the travel phase of the slide-out, then re-engage the pins at the conclusion of the motion. This would occur at either direction of the motion; extension or retraction.

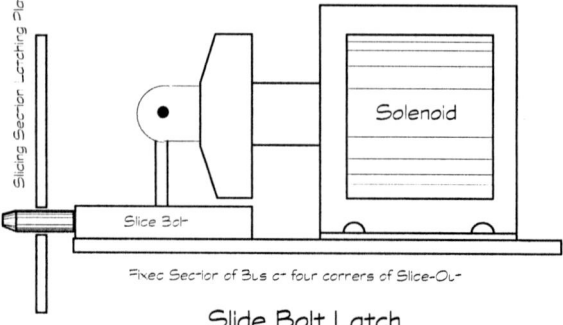

Slide Bolt Latch

Finally, an extremely simple form of latch is to use the air cylinder actuated by the skinner valve to be a slide pin. No doubt a number of this springloaded air cylinder may be

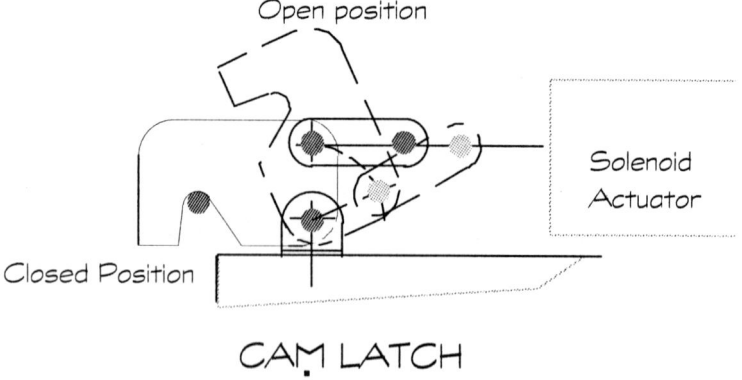

CAM LATCH

scrounged at any truck wrecking yard. The cylinder is springloaded in the extended position. By applying air pressure, it retracts. In it's motor application, it's retraction allows the rack to open delivering fuel to the injectors. As a latch application, the retraction need only occur during the slide movement.

Shown is a Detroit diesel shut down cylinder controlled by the skinner valve. Spring-loaded extended. The skinner valve applies pressure to retract the plunger, thus opening the rack on the motor allowing fuel to flow to the injectors. This unit may be adapted to a pin latching devise where the electrical controls to extend the slide-out room will open a skinner valve to retract the cylinder during the transition phase in either direction.

For simplicity, you can always use a set of bolts and nuts, or even some "C" clamps (just kidding!)

Slide-out rooms
mechanics and
Structures

Dave Galey

Electrical Controls

As far as the electrical controls, a 120 VAC system would have more parts readily available. The simplest trick is to purchase a chain driven electrical garage door opener and modify the system to operate your slide-out room.

One of the beauties of the garage door opener is it come complete with push button, or remote switches along with limit switches. Plus they have a built-in clutch to over ride the extension or retraction in the event an interference is met.

The sequencing I envision is to push the button to extend or retract at which point the solenoid latches disengage during the travel sequence. At the completion of the travel sequence, a limit switch disengages the power to both the motor and the solenoid and a spring re-engages the slide bolts latches.

The diagram on page 81 shows the control switch spring-loaded to the Center Off Position. To extend the slide-out section, the switch moves to position 2, allowing power to flow through the *Limit Switch out* to position 6 to position 7, turning the motor clockwise. The current returns from terminal 9 through 10 to 14 to ground. While power is applied, the latches are disengaged, to return to engagement when power is off. When the slide-out section reaches it's extended position, it opens the limit switch, disengaging power.

Slide-out rooms, mechanics and Structures

To retract the section, the control switch moves to position 3. This allows power to flow through *Limit Switch In* actuating the DPDT Relay, delivering positive current to teminal 9, reversing the motor, thus retracting the section. Again, the power is shut off when the unit moves to trip the Limit Switch to the In position.

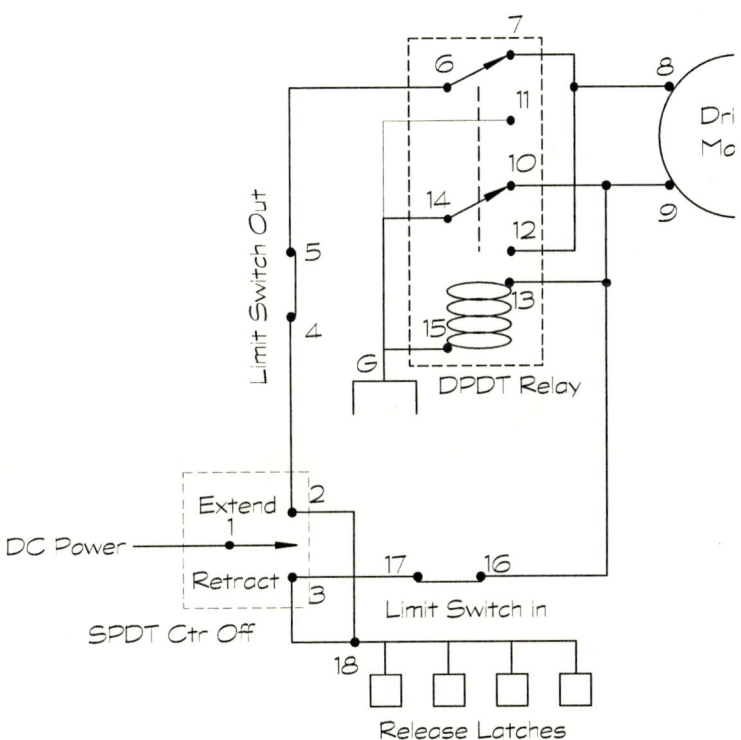

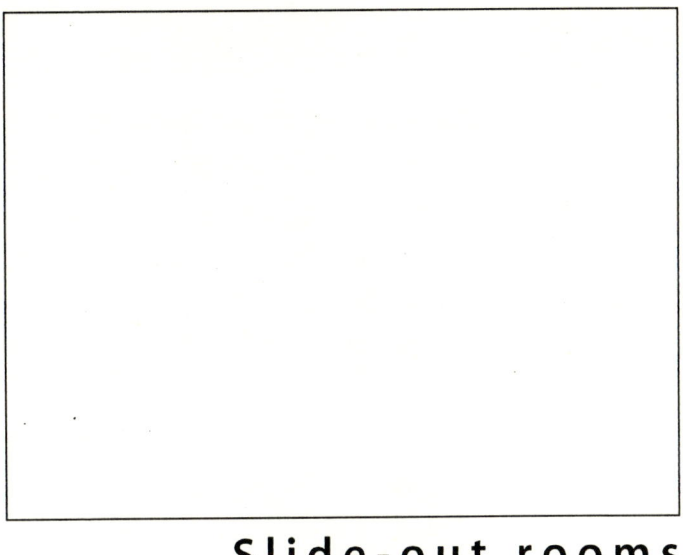

Slide-out rooms mechanics and Structures

Dave Galey

Conclusions

The theory, concepts and drawings in this presentation are some ideas on how to accomplish the goals set forth. The reinforcement approach is radically different from one we took several years ago. In this approach, an effort was made to preserve a good percentage of access area to the baggage compartment. About all this proves is, there are many *ways to skin a cat*.

The ideas shown are the authors ideas and, no doubt, better ideas will be developed. If this were not true, we would all still be driving Model "Ts".

The exciting thing is, we keep *pushing the envelope* and this has literal meaning when consider slide-out rooms. It is my hope this book will stimulate some self conversion specialists to do some more pushing. I will welcome any suggestions and criticisms to add to this small book.

Good Luck! and
Have Fun!

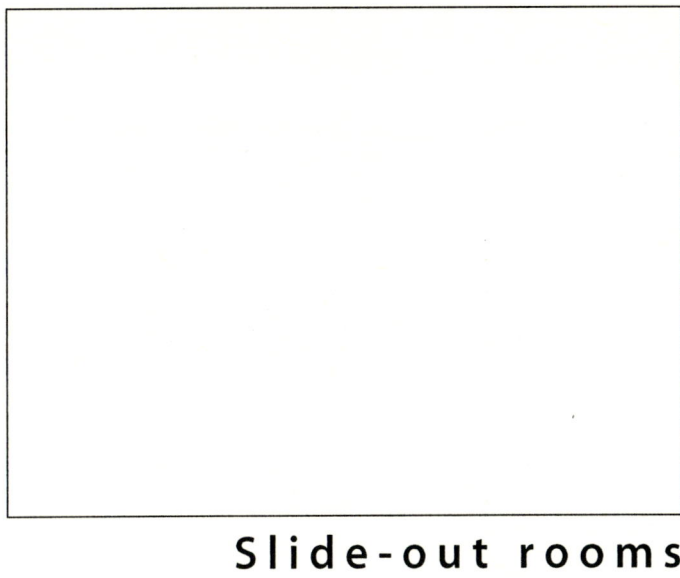

Slide-out rooms mechanics and Structures

Appendix

The following pages contain a series of representative drawings which may be adapted to your design and may be modified to suit your needs.

Every effort has been made to be generic and not favor one bus manufacturer over another. The drawings are suggestions only. Each application must of necessity be unique

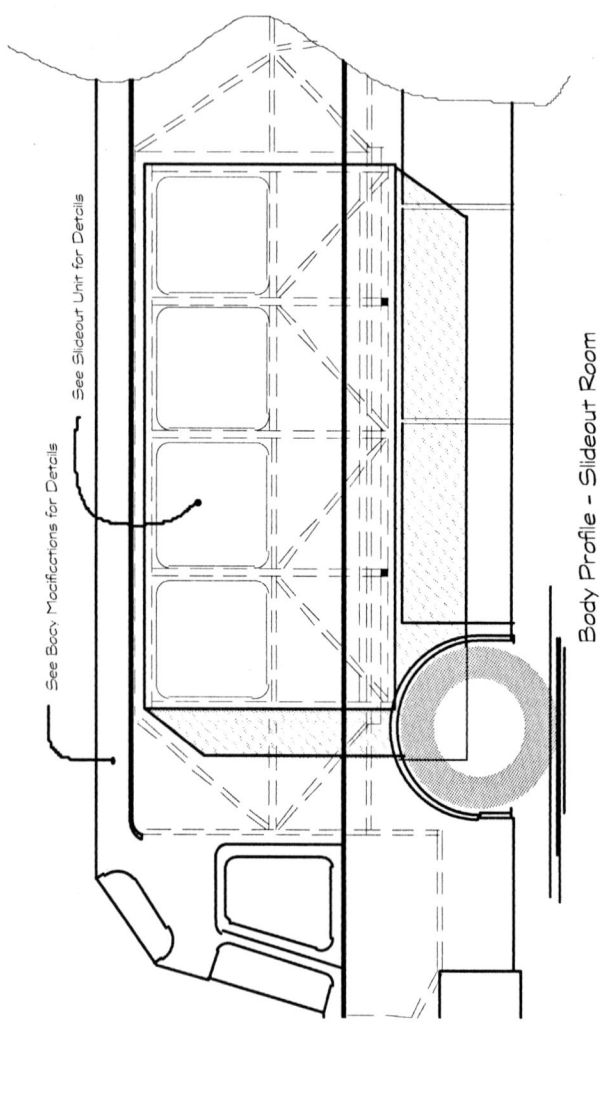

See Slideout Unit for Details

See Body Modifications for Details

Body Profile - Slideout Room

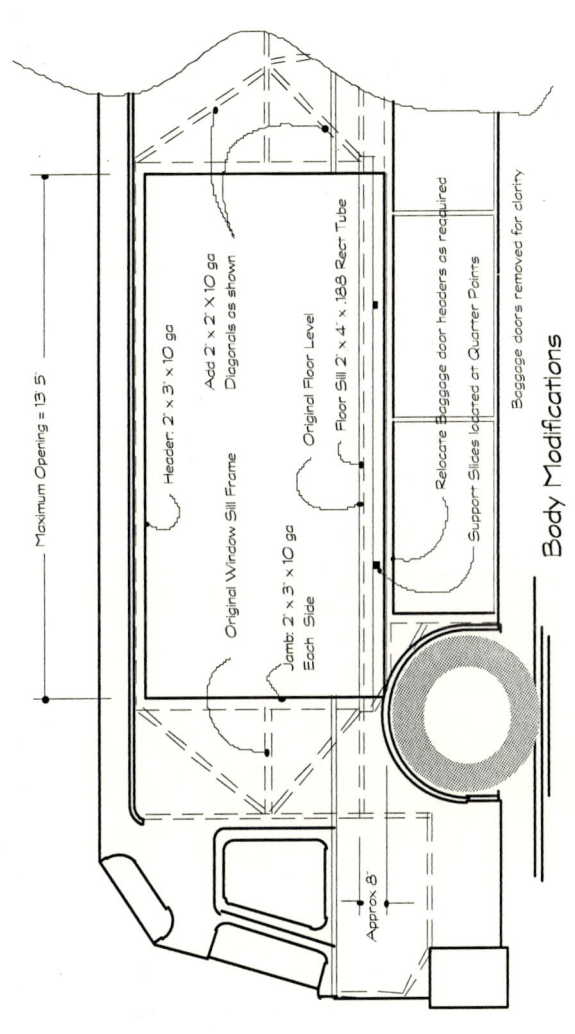

Maximum Opening = 13' 5"

Header: 2" x 3" x 10 ga

Add 2" x 2" X 10 ga
Diagonals as shown

Original Floor Level

Original Window Sill Frame

Floor Sill 2" x 4" x .188 Rect Tube

Jamb: 2" x 3" x 10 ga
Each Side

Relocate Baggage door headers as required

Support Slides located at Quarter Points

Baggage doors removed for clarity

Approx. 8"

Body Modifications

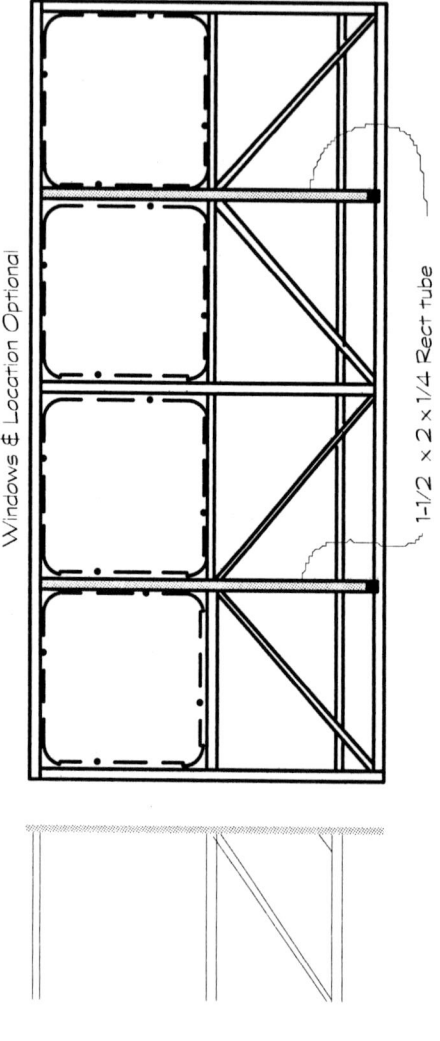

Windows & Location Optional

1-1/2 × 2 × 1/4 Rect tube

Balance of framework 1-1/2 × 1-1/2 × 1/8 Sq tube

Slideout Unit

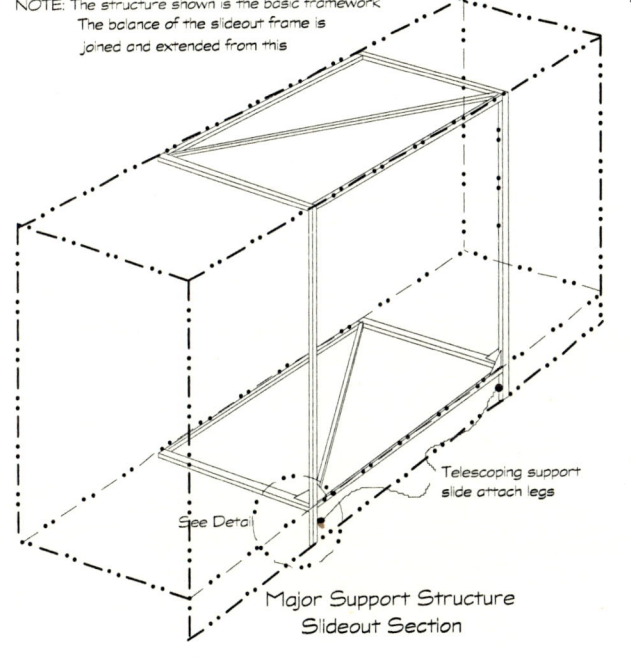

NOTE: The structure shown is the basic framework
The balance of the slideout frame is
joined and extended from this

See Detail

Telescoping support
slide attach legs

Major Support Structure
Slideout Section

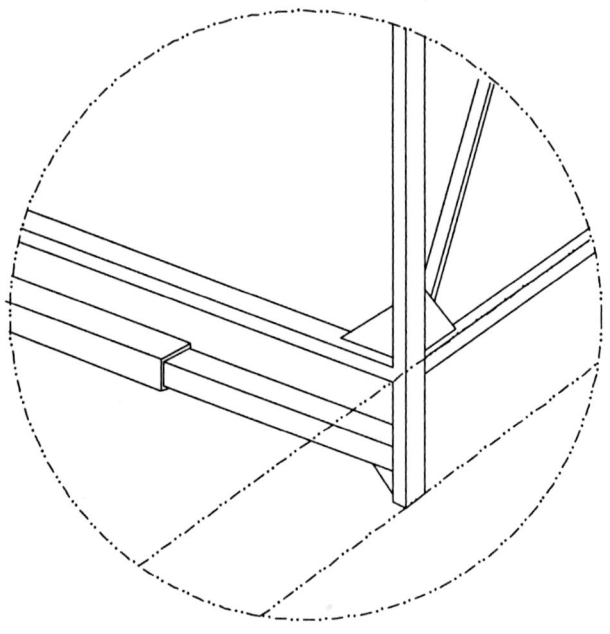

Corner Detail showing telescoping
slide attached to Slideout section

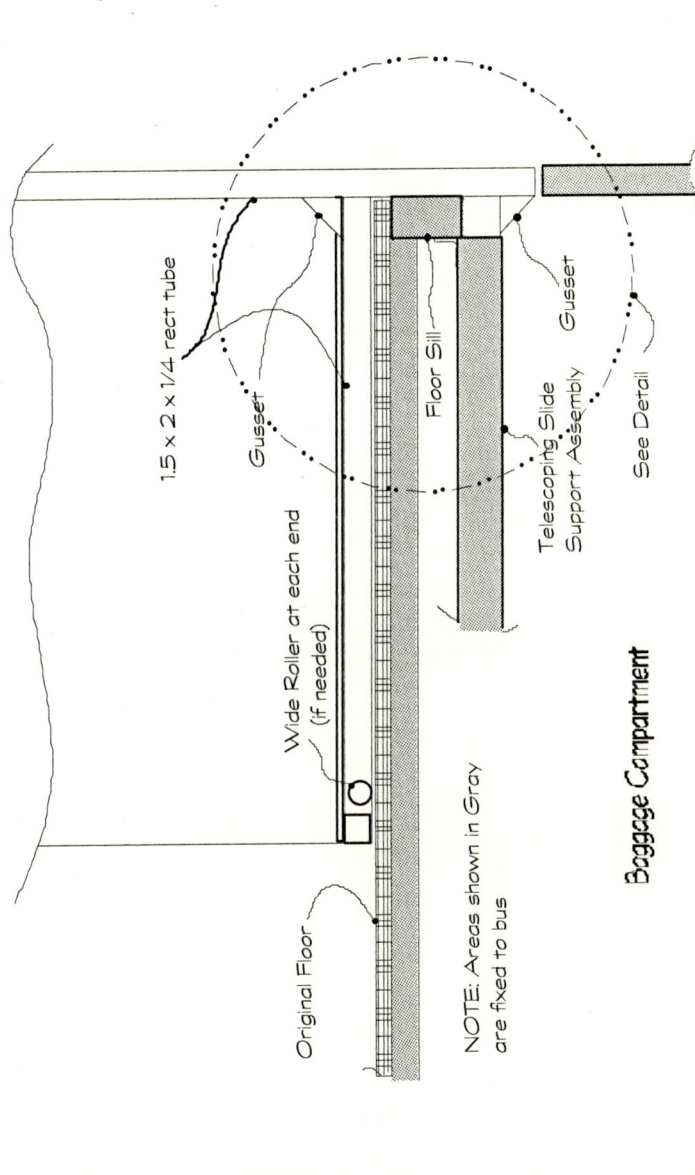

1.5 x 2 x 1/4 rect tube

Gusset

Wide Roller at each end
(If needed)

Floor Sill

Telescoping Slide
Support Assembly

Gusset

See Detail

Original Floor

NOTE: Areas shown in Gray
are fixed to bus

Baggage Compartment

Slide-out rooms, mechanics and Structures

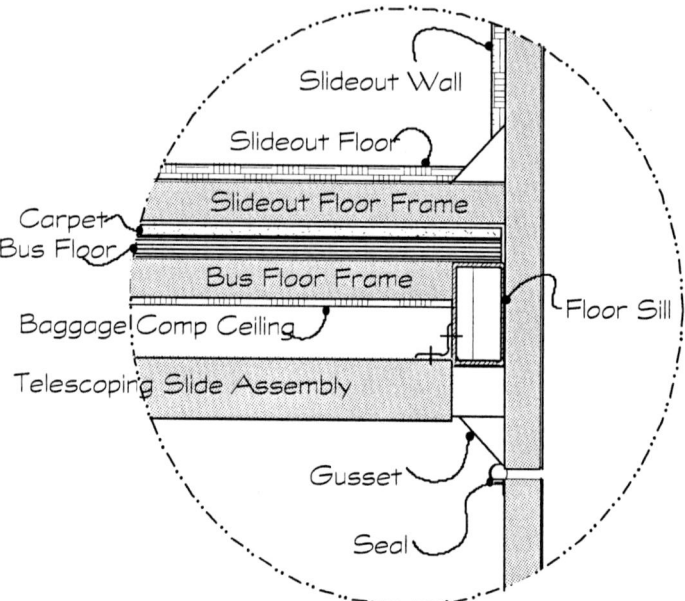

Detail from previous page